Manuel Francisco Bandeira

Tourism potential of Dombe Grande, Baía Farta, Benguela in Angola

Manuel Francisco Bandeira

Tourism potential of Dombe Grande, Baía Farta, Benguela in Angola

Natural tourist destination and mystical culture

ScienciaScripts

Imprint

Any brand names and product names mentioned in this book are subject to trademark, brand or patent protection and are trademarks or registered trademarks of their respective holders. The use of brand names, product names, common names, trade names, product descriptions etc. even without a particular marking in this work is in no way to be construed to mean that such names may be regarded as unrestricted in respect of trademark and brand protection legislation and could thus be used by anyone.

Cover image: www.ingimage.com

This book is a translation from the original published under ISBN 978-613-9-61471-4.

Publisher:
Sciencia Scripts
is a trademark of
Dodo Books Indian Ocean Ltd. and OmniScriptum S.R.L publishing group

120 High Road, East Finchley, London, N2 9ED, United Kingdom
Str. Armeneasca 28/1, office 1, Chisinau MD-2012, Republic of Moldova, Europe
Printed at: see last page
ISBN: 978-620-7-63098-1

Table of contents:

PROVINCE OF BENGUELA KIBABATA SERVICE-LDA CONSULTANT

POTENTIAL OF DOMBE GRANDE COMMUNE, BAÍA FARTA MUNICIPALITY

Manuel Francisco Bandeira

DOMBE GRANDE, 2018

THE TOURIST POTENTIAL OF DOMBE GRANDE, MUNICIPALITY OF BAÍA FARTA, BENGUELA PROVINCE IN ANGOLA Manuel Francisco Bandeira

Technical team:
Ivan Guardado de Sousa Inácio Gil
Tomás Justino Guia

THOUGHTS

"In order to achieve harmony and forest ecotourism, it is urgent to combat the already growing *ecotourism* in our country[1].
Egotourism, rather than ecotourism, does not have the aim of preserving the environment, but rather the selfish desire for satisfaction and narrow recreation that does not respect the laws of nature, the principles of civility and coexistence, and seeks to negatively alter the environment."

Tati, J. A. (1999:14)

"The historical sites, the parks, the landscapes, the rivers and valleys, the mountains, the sea and even the rocks, if well used, are products of tourist attraction."

Armando da Cruz Neto (Benguela Magazine, No. 3, November, 2011:11).

---

[1]*Egotourism: a* neologism used by the author to differentiate sustainable tourism from unsustainable (depredatory) tourism; a practice that tends to increase in some places in Angola. In: TATI, J. A. (1999): "Ecoturismo, importancia e perspectivas de desenvolvimento em Angola".*Angola Turística*, 3: 14-18, May/June, Luanda.

DEDICATORY

To the Mundombe people for allowing us to carry out this work in the Dombe Grande commune.

ACKNOWLEDGMENTS

To the Communal Administration of Dombe Grande, for the trust granted to carry out this work;

To the Technical Team for Local Development (ETDL) of the Department of Natural Sciences (DCN) of ISCED Benguela of the Katyavala Bwila University (UKB), for their support and technical-scientific advice;

To all those who accompanied us on the fieldwork (geographical excursion to identify and record resources and places), especially the Communal Administrator of Dombe Grande, Mr. Bertolino Asdruba Soares Morais;

To all the authorities who supported us, especially Soba Gongalves Paulo for his wisdom and Mr. Martinho Lopes (responsible for culture) for his help and support;

We would like to thank everyone who collaborated directly or indirectly in collecting and providing the data to make this work a reality.

PREFACE

I was honored to be invited by Professor Manuel Francisco Bandeira, the main author of the book entitled: **"Tourism Potential of the Dombe Grande Commune, Baía Farta Municipality, Benguela", a** task which I timidly accepted, for two fundamental reasons. Firstly, because the main author is special to me and whom I greatly admire as a colleague, friend and former teacher, and secondly, because it is a subject that I personally have a deep and particular passion for, because it is a subject that is much talked about in Angola, but about which little is known and done, even though it actually represents the new wealth of success and economic and social development for countries that have been able to understand its potential in time.

In fact, at a time when countries are scrambling to find solutions to counteract the direct and indirect effects of the global economic and financial crisis, with all its consequences for the real economy, the definition of economic, income-generating alternatives - outside the traditional profiles of their productive and commercial activities - is a strategic imperative, is a strategic imperative and an action that should be encouraged by governments in their public policies, with a view to creating and boosting new activities capable of contributing to the production and increase of national wealth, enabling them to diversify their economies and, through this, reduce the high levels of poverty that plague these countries of which Angola is a part.

Tourism has evolved significantly in recent years and the impact it has had on socio-economic sectors has led countries and international institutions to pay particular attention to it. A number of scholars have even stated that the governments of developing countries, particularly those in Africa, would probably have achieved better rates of economic and social development if they had invested in a committed way in their development strategies to boost tourism resources, This would create positive externalities for the dynamization of other traditionally non-exportable sectors, such as agriculture, livestock, trade, culture and transport, thus enabling the services of these economic segments to become "invisibly exportable" goods.

This framework would therefore not only create benefits for broadening the employment base of intensive labor, which in these countries is extensive due to the high levels of illiteracy and limited opportunities for schooling of the economically active population - due to a lack of schools, teachers and good policies - but would also allow governments to better redirect their policies towards the development of traditional sectors that are economically competitive on the international market, such as mining and oil.

Tourism is currently the "*transmission belt*" between North and South and is of interest to everyone - industrialized and developing countries alike - being a truly sustainable alternative to all other segments of economic and productive life, contributing above all to the conservation of the environment, the improvement of the well-being of local communities, security, the deepening of human friendship and solidarity - values which are some of the flagships of the defense of human rights.

In the case of Angola, and particularly for the province of Benguela, the development of tourism is an option that the government should not discount precisely because this province has a natural ecosystem that can turn it into the country's biggest tourism development hub. Benguela has beaches, the sea, semi-desert, mountains, cliffs, forests, fauna and flora, a unique and diverse cuisine, but also the nuances of its people, dangas and cheerful, hospitable people.

The book deals with the tourist potential of Dombe Grande, which is also a municipality whose historical and cultural weight in the mosaic of the province's development has made it an unavoidable reference point, due to its monuments, its traditions expressed in the dangas, its initiation events "Efiko and Ekwedje", the feitigo, but also its beautiful beaches and happy people. And the fact that this town is located on the most economically and socially dynamic coastal corridor in the country, linking Benguela, with its four cities (Lobito, Catumbela, Benguela and Baia-Farta) with the province of Namibe, makes it the gateway to the emerging metropolitan area in a south-north direction, with a population concentration that could soon reach 3 million people. That's why tourism could become one of the main axes and springs for the development of this town and the municipality of Baía Farta in general.

The main author, who is probably one of the best tourism specialists and scholars in Angola, has analyzed the material and immaterial resources of this important and legendary town of Dombe Grande with the thoroughness, rigour and sensitivity that characterizes him, and this, after having done similar work with several other provinces in Angola, which necessarily leads the dedicated institutions of the Government of Angola and the Province of Benguela in particular, to view this study and contribution by this Angolan specialist with the attention required and the necessary support.

I therefore invite you to read this work because it is Angolan and it is for Angola.

BRAGA, APRIL 18, 2018
JOSÉ JANUÁRIO

INTRODUCTION

Tourism in rural areas is a form of activity that prioritizes leisure and recreation focused on the natural landscapes, cultural heritage and local development of inland regions. One of the main objectives of this activity is to promote an integrated understanding of the environment in its multiple and complex relationships, involving natural, social, economic, cultural and ethical aspects. *Rural tourism* can be included in the category of what is known as natural and/or exotic tourism; a combination of agri- and ecotourism, with an emphasis on enhancing regional cultural identity and improving the living conditions of the local community.

From this perspective, tourism can become one of the main strategic axes for local development in the Dombe Grande commune and play an important role in the process of diversifying the province's economy, as tourism can perfectly complement the agricultural and livestock sectors that predominate in the territory, to the extent that the commune has a diverse range of valuable *natural, historical, monumental and cultural* tourist resources that need to be properly identified, systematized and classified for subsequent planning and to be placed in the tourism network of Benguela in particular and of the country in general.

It should be emphasized that, in addition to the inventory and diagnosis, it is important to draw up a program for the dissemination, promotion and marketing of the Dombe Grande commune's tourist resources to serve as a starting point for making the appropriate decisions to promote the necessary investments and activate the harmonious and rational planning of tourism at local level, taking into account the pillars of sustainable tourism.

The book is structured in three chapters, the most important of which is the *Introduction,* which gives a brief overview of the inventory, presenting the general and specific objectives and the methodology used to support the theoretical and practical aspects of the work. The *first chapter* characterizes the physical-geographical, economic and socio-demographic aspects of the territory under study. The *second chapter* presents a systematized record and identification of the main tourist resources and attractions in the commune, according to a previous classification. The *third chapter* presents some proposals and strategies for action, reflecting in part the approach to the analysis of Weaknesses, Threats, Strengths and Opportunities (DAFO), internationally known as SWOT analysis, to support these proposals and implement tourism activities in the Commune in question. Finally, the conclusions, some recommendations, annexes and other related structural aspects that highlight the final result of the work are presented.

CHAPTER I

PHYSICAL, ECONOMIC AND GEOGRAPHICAL ASPECTS OF DOMBE GRANDE COMMUNE

1.1. Geographical location and territorial limits

The Dombe Grande commune is located to the south of the seat of the municipality of Baia Farta, at a distance of just over sixty (60) kilometers. It is situated at Latitudes 12° 45' North and 13° 03' South and between Longitudes 12° 55' West and 13° 13' East. Its territorial extension is 2,172.2 km^2 , corresponding to 32.2% of the territory of the Municipality of Baia Farta. The Commune is subdivided into 13 (Thirteen) Bairros, 12 (Twelve) villages and 7 (Seven) Povoagóes. Its territorial limits (borders) are as follows:

- *North and Northeast*: Baia Farta headquarters;
- *East*: Kalohanga Commune;
- *Southeast*: Bolonguera commune (Chongoroi municipality);
- *South*: Kamukuio Municipality (Namibe Province);
- *Southwest*: Equimina Commune;
- *West*: Atlantic Ocean.

Figure 1: Map of Dombe Grande Commune

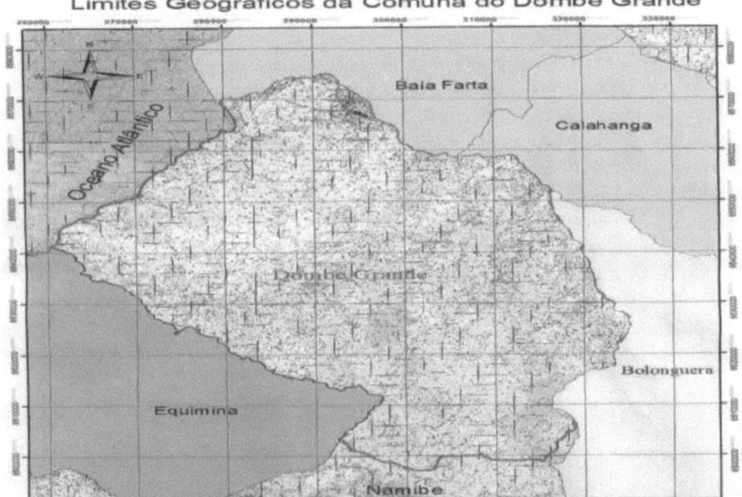

Source: Prepared from IGCA - Benguela (2013).

1.2. A brief history of the Dombe Grande commune

The town of Dombe Grande[2] is not very old. Reports say that its people date back to before the arrival of the Bantu peoples. The natives are generically known as *Mundombe* and were the protagonists of a unique moment in our history. Tradition has it that in the hills of Dombe Grande, the people of the region defeated the Jagas, warriors from Central Africa who, in the 16th century, spread terror throughout what is now Angola. After wiping out the all-powerful Kingdom of Congo, the Jagas headed south. When they reached the hills of Dombe Grande, they were finally defeated by the Mundombe, who put an end to the invaders' reign of terror. The story is still passed down by word of mouth among the elders.

---

[2] **Dombe Grande** was also called Dombe da Quizama or Dombe da Quinzama. Relations between the people of Dombe and the Portuguese authorities date back to the beginning of Cerveira Pereira's occupation of Benguela in 1617, although they were not always friendly. A year after Cerveira Pereira's installation, when the soba of Quinzama refused to pay him vassalage, he forcibly subdued him, thus making him a faithful ally (In Alexandra Aparicio).

However, during the centuries that followed, the Mundombe were unable to resist the greatest threat of all: slavery. The fate of thousands of people was tragic, but their mark reached far and wide! Several historians say that the Mundombe or "Ndombe" gave their name to a strong tradition that still celebrates African heritage in suspected Uruguay[3] .

The resistance of the natives is linked to the figure of Vaita, the king of the Ndombe, who at the time protested to the colonial authority against what he considered to be injustices to which his people were subjected, especially the high taxes and the mistreatment of the natives. In response, he was arrested, which infuriated the population, who demanded the king's freedom for days. This happened on August 18, 1856, giving rise to the traditional Ndombe festivities, which used to last several days.

1.3. Relief and geomorphology

The territory is part of the coastal plain that extends in a north-southeast direction. Scattered mountains can be seen in the area, including the Abelhas, Vihombo, Cacorongo and Chipupa mountains, with varying altitudes averaging 423 meters. The vast territory also includes depressions (Cuio Chamume, 2-7 meters) and extensive valleys dominated by the Coporolo River. In summary, the relief is of the undulating, denuded-accumulative type interspersed with elevations, river valleys and open valleys.

1.4. Climatic aspects and hydrography

The *climate is tropical dry (hot and semi-desert)*, influenced to a large extent by the proximity of the sea, such as Cuio beach, located 12 kilometers from the municipal seat. The average annual temperature is 23.7 degrees, the average annual rainfall is 120 millimeters and the relative humidity is 72.7%. Rainfall varies from the coast to the interior, but the vegetation cover around the headquarters has a positive influence, facilitating agriculture in the outlying fields. In an easterly direction (inland), the climate is somewhat milder due to the continental factor. In addition to this, there are the wide valleys flooded by the Coporolo river. With regard to the *hydrographic network in the* commune, the *Coporolo* River, which rises in the municipality of Chongoroi and is closely linked to the history, identity and mystique of the commune, the Tchitandalúcua River and the Calongue River, which merge to form an estuary and a lagoon at their mouths, stand out.

1.5. Characteristics of flora and fauna

The predominant vegetation (flora) is of the steppe type, semi-desert combined with shrubs of the hawthorn and acacia species (Banje Kumbi) scattered from north to south. Some xerophytic species of the cataceae family and normally dry grasses can be seen. In the middle of the town, there is an anthropogenic vegetation cover, influenced by the peripheral, cultivated fields and leafy hundred-year-old trees, making the town's environment pleasant for visitors.

As for the associated *fauna*, rabbits, gazelles and fanged goats stand out, mainly in the extreme south and southeast of the territory, where the Rodrigues caga area stands out.

1.6. Socio-demographic and economic aspects

The Dombe Grande Commune[4] has an estimated population of 41,434 inhabitants[5] , 19,345 male (46.75%), 22,089 female (53.3%), with a population density of 19.07 inhabitants/km^2 made up of people from the *Ndombe, Kwisi, Ovakwando, Ovimbundu, Ngangela and Muhumbi ethnolinguistic groups*. Although some scholars group the Ndombe people together in the Ovimbundu ethno-cultural area[6] , their characteristics allow them to be grouped together in the Ovahelelo ethno-cultural area (Gomes 1999: 33). Although the Ndombe are the original peoples of this region, today it is possible to identify some representatives of ethnic groups such as the Ambundu, Bakhongo and Lunda-Cokwe who have settled in the region for various reasons.

3Candombe, as it is called, is a rhythm based on drumming that originated with the slaves who arrived in the Rio del Plata region from the 15th century onwards. Today, it is an artistic expression and an affirmation of the Afro-descendant community of this small South American country, which knows that its "candombe" is linked by an umbilical cord to the "Dombe Grande of Angola". A story almost unknown to us, but celebrated with pomp and circumstance in Uruguay, which succeeded in having candombe declared a World Cultural Heritage Site in 2009.

[4]Data obtained from the local administration.

[5]General population and housing census (2014).

[6]In Fernandes et Ntondo (2002: 54).

The religiosity of its populations is manifested through various faiths represented mainly by the Catholic Church, the Evangelical Congregational Church of Angola, the Evangelical Synod Church of Angola, the Universal Church of the Kingdom of God, the Church of Our Lord Jesus Christ in the World (Tocoists), and the Pentecostal Assembly of God, among others, in addition to local religions, making a total of 24 religious denominations recognized by the government[7] .

The education system is functional, with just over 29 schools, 26 of which are primary schools and the rest secondary schools. At this level of education, there is a 2nd cycle dedicated to teacher training. In 2017, the teaching staff consisted of 389 teachers, 235 of whom were in primary education, 101 in the 1st cycle and 53 in the 2nd cycle of secondary education.

During this period, the student population was 14,856, broken down as follows: a) Primary Education: 12,259 students (of which 1,658 were in adult education); b) 1st Cycle of Secondary Education: 2,799 students; c) 2nd Cycle of Secondary Education: 1,327 students.

Table 1: Student and teaching staff in Dombe Grande Commune

Level of Education	Student body		Teaching staff	
	MF	F	MF	F
Primary	12.259	5.825	235	118
1st Cycle	2.799	1.227	101	31
2nd Cycle	1.327	631	53	15
TOTAL	16.085	7.021	389	164

Source: Own elaboration

Table 2: Dombe Grande Commune teaching staff qualifications[8]

Level of Teaching	Teaching staff		Academic level of the teaching staff													
			Tech. Basic.		Tech. Doctor.		Frequency Univ.[9]		Lic		Postgra		Master		Doctor	
	MF	F	MF	F	MF	F	MF	F	MF	F	MF	F	MF	F	MF	F

---

[7] Annual report (2017).

[8] Source: Dombe Grande Commune School Coordination, 2017; Santa Ana Missionary School Complex, 2017.

[9] This category includes teachers who are still studying at university and those who have not yet presented and defended their monograph. In the original document they are categorized as Bachelor, 3rd and 4th Year respectively.

Primary	235	118	14	7	161	80	48	28	31	8	-	-	-	-	-	-
1st Cycle	101	31	-	-	40	14	30	18	31	8	-	-	-	-	-	-
2nd Cycle	53	15	-	-	3	0	9	3	28	9	2	-	7	1	1	1
TOTAL	389	164	14	7	204	94	87	49	90	25	2	--	7	1	1	1

Source: Own elaboration

Table 3: Distribution of schools by location

Location	No. of schools		
	Primary	1st Cycle	2nd Cycle
A good reminder	3	110	-
Luacho	5	1	-
Dombe Grande (Headquarters)	15	2	1
Singing	1	-	-
Kambonge English	1	-	-

Kangengo	1	-	-
TOTAL	**26**	**3**	**1**

Source: Own elaboration.

Dombe Grande's health system, especially at the headquarters, is fundamentally supported by the Communal Hospital, the Kalundu Health Center and 4 Health Posts. The commune has 86 nurses, who work at the hospital and at eight (8) medical posts and at the Maternal and Child Center located at the headquarters. There is a senior technician, three (3) general practitioners, two (2) nationals and one (1) foreigner of Korean nationality. Part of the population uses traditional medicine (healers).

With regard to drinking water supply, there is a new system in the commune for collecting, treating and distributing water in the village of Kangengo, but it is below the real needs of the local population. Household energy is deficient, and there is support for those served by the Communal Administration and a few houses. However, there are two generators for public lighting, only one of which is in operation.

Agriculture is the main economic activity, with maize, manioc, sweet potato, kidney potato, beans, vegetables (tomatoes, peppers, onions,[10] cabbage, kale), fruit (bananas, mangoes, guava, papaya, guava). This is followed by *livestock farming,* with cattle and goats. *Fishing* involves drying and salting fish. The industrial sector had Agucareira 4 de Fevereiro, the driving force behind the region's development, until it was completely shut down in 1991. There are small-scale functional factories dedicated to bread production, namely two (2) bakeries and eight (8) mills, a metal packaging factory (cans) and a tomato processing plant. Formal commerce[11] is represented by a few stores and warehouses, and the informal sector is represented by small pests that complement the productive sector. The tourism sector is practically incipient, represented by a few units[11 12 13] .

A series of farmers' cooperatives and associations have been promoted through the intervention of national and foreign NGOs. It was within this framework that the NRA[13] (Núcleo Representativo das Associagoes do Dombe Grande) emerged, which brings together 14 peasant associations, service providers and 2 cooperatives of small agricultural producers, and CAPIAD[14] (Cooperativa Agro-Pecuária e Industrial do Dombe Grande), which brings together 78 medium and large agricultural entrepreneurs.

---

[10] Rooms attached to the Sapangele 1st Cycle School, located in Povoagao do Luacho.

[11] There are 350 market traders, 14 stores and 58 canteens in the Commune.

[12] One (1) inn, one (1) esplanade, one (1) restaurant bar, one (1) lodge, 7 snack bars, 31 taverns and one (1) roulotte.

[13] The NRA is the result of the community intervention work carried out by ADRA-Antena Benguela between 1993 and 2005.

[14] CAPIAD came about as a result of CLUSA's work to support agricultural entrepreneurs through the implementation of PROAGRO - the Financial and Agricultural Development Program, between 2006 and 2012.

CHAPTER II

LOCAL TOURIST POTENTIAL
2.1. TOURIST POTENTIAL OF THE HEADQUARTERS AND OUTSKIRTS
2.1.1. Natural Tourist Resources
4- Chimalavera Regional Park

<u>Date of establishment</u>: It was established as a special reserve by Legislative Order Number 352 of April 15, 1974.

<u>Geographical location</u>:

S Geographical boundaries: 12° 45' to 12 16' South Latitude and 110 44' East Longitude.

S Nearest city: Benguela, 42 km by road.

<u>Physiography</u>:

S Land area: 150 km^2

S Altitude: 50 meters; maximum 262 meters.

S Topographical features: high plains surrounded by rugged mountains.

<u>Hydrographic network</u>: Existence of rivers, lagoons and marshes.

<u>Climatic characteristics</u>:

S Temperature: Annual average (23.5° C); coldest month - July (19.4° C), warmest month - March (26.8).

Precipitation: Annual 305mm; precipitation for 45 days.

S <u>Humidity</u>: Annual average 77%.

S <u>Typical flora</u>: Sub-littoral steppe formations of *acacia mellifera sp, ditinens and acacia sp.*

S <u>Predominant fauna (species)</u>:

- Large mammals: Grey monkey, striped jackal (*Canis adustus*), fan goat (*Antidorcas marsupial*), and olongos (*Strepsiceros strepsiceros*) - Predominant species: Fan goat.

- Species of special importance: Fan goat.

<u>Anthropogenic impact</u>: The prospect of transforming the park into a tourist attraction, capable of continually increasing visits and the occupancy rate of the bungalows.

<u>Tourist infrastructure</u>: The park has tourist infrastructure comprising eight (8) bungalows with the essential conditions for tourist accommodation (air conditioning and DSTV signal) to guarantee an optimum stay. The access road is in good condition and the park is managed by Imogestin.

Proposal:

Raising investment with a view to incorporating some species of animals that are more attractive and adaptable to the area, promoting campaigns to publicize, promote and market to specific segments of demand in order to attract more tourists and internationalize the area.

Images: View of access to the bungalows, reception-restaurant and sample of bungalows to accommodate tourists.

-I- Chipupa Spring

It is a spring located at the geographical coordinates: Latitude South: 12° 45' 068" and Longitude East: 013° 13' 182". The spring is located at the base of the hillside near one of the sides of the Benguela-Dombe Grande road. According to physico-chemical analyses, the source discovered has medicinal mineral characteristics and can be used for therapeutic purposes (health tourism).

Proposal:

The feasibility of installing a restoration infrastructure, making rational use of the water source for tourist purposes, taking advantage of the therapeutic purpose of the water for health tourism (spa) could be studied.

Images: View of the surrounding landscape and the place where water is drawn from the spring.

4- Water gates (Coporolo River)

It is located to the east of the headquarters, just over 8 minutes away by car. Geographical coordinates South: 12° 59' 11.7" and East: 13° 07' 77.5". It is located on the left bank of the river, where there used to be a machine to control the water pipeline that supplied the 1° de Maio sugar factory. There is a fluvial beach on the left bank of the river, creating an ideal spot for sunbathing and a beach in the middle of the river. Underused to a certain extent, it serves as a leisure space, especially for young people who seek tranquillity there, and it also serves as a place for religious camps.

It is an anthropo-natural landscape that highlights a certain landscape uniqueness; a combination of physical-geographical and geological elements: the canyon over the Coporolo River, the gallery-shaped vegetation combined with medium-sized trees and a variety of fauna integrated with birds (several species of birds) and some monkeys visible mainly on the left bank. The environmental status of the site is reasonable, but there has been uncontrolled dumping of solid waste (cans and plastics).

Proposal:

The site offers natural conditions for the construction of a youth camping infrastructure. The idea could make the site more attractive and profitable (employment and income). The installation of an eco-camp aims to turn the site into a reference attraction. To this end, a public-private partnership could be used (administration, Teacher Training School plus two or three local entrepreneurs (to design a specific project).

Images: Views of the meandering Coporolo River and Canyon from the top.

-1- Bichor Viewpoint

An interesting spot for observation, located to the south of the headquarters in the direction of Namibe, on one of the hills situated on the way up from Bichor at the geographical coordinates Latitude South: 13° 59' 573" and Longitude South: 13° 06' 024". From the site, you can see almost all of the town, forming a magnificent landscape that includes the town, the outlying areas, the green space drained by the river (crops), the river itself with its sinuosity, the bridge and the riverside communities that surround it and have the river as their source of life.

Proposal:

Build a spire on the site to ensure that tourists can stay and observe and contemplate the town's landscape. You can use rustic construction with local materials such as stones, sticks (bamboo) and trees to create an environment that fits in with the natural surroundings.

Image: From the viewpoint, you can see the shape of the town of Dome on the distant horizon. Big.

-2- Rodrigo's shit area

This is a relatively vast area, located to the southeast of the municipal seat in the direction of Equimina at geographical coordinates: South Latitude: 13° 06'119 and East Longitude: 13° 04' 384" at an altitude of 367 meters. It is a lowland area with shrubby vegetation and creeping grass, typical of the steppe formation, which is somewhat affected by rainfall. The area is home to several species of animal, such as the fan goat, the black monkey, rabbits, impalas, rodents and birds, among other species that make it the preferred area for small and medium-sized game.

Proposal:

Delimit the hunting area and create a mechanism to control the activity in order to protect the species. Try to keep a record of lovers of this activity with a view to activating ecotourism, as it may be possible to determine observation zones at specific times of the year.

Image: View of the landscape in the Rodrigo area, showing the bushes and the absence of creeping grass due to the lack of rain at the time.

2.2. CUIO'S TOURIST POTENTIAL

2.2.1. Natural Tourist Resources

Dombe Grande's main beaches

The main beaches of Dombe Grande are located to the south of the town of Cuio. In general, they are located between steep limestone hills (cliffs), in which water lines flow forming sandy coastlines, with favorable natural conditions for *sun and beach tourism* and a range of related activities, including sports tourism (fishing, canoeing, submarining and diving) and other related activities.

The beaches identified are as follows:

4- Cuio Beach

Location: Located to the SW of the town hall.

Geographical coordinates:

Z <u>Far North</u>: Latitude South: 12° 55' 860"- Longitude East: 12° 57' 801".

Z <u>Far South</u>: Latitude South -12° 59' 06.4"- Longitude East: 12° 58' 29.5".

Length: Approximately 5,600 meters

Width: Approximately 50 meters in the widest part (towards the mouth of the Colocolo River).

Type of beach in terms of use: Non-urban beach with intensive use, due to the area's high population density related to artisanal fishing activity.

Type of beach in terms of slope: Intermediate type at the southern end and tombolo at the southern end.

Size and color of the sand: Fine brown sand.

Visitor turnout: Not high. Most are concentrated in the far south. Adventurers with 4X4 vehicles head north to the mouth of the Coporolo River.

Access routes: Cuio beach can be reached by land, along a 15-kilometer stretch of dirt road on rocky and rugged terrain. The road should be rehabilitated regularly to facilitate access.

Environmental status: Reasonable in the far south. However, it is necessary to improve the organization of the huts and fish drying stations, and to take care in their management (hygiene), as solid waste (abandoned cans and irons) is present.

Degree of occupation of the peripheral space: Presence of the Pescuio fishery infrastructure, which was a large-scale enterprise in the past, but is currently dedicated to marketing the fish supplied by artisanal fishermen, as it does not have the vessels to produce on a large scale. A large-scale agricultural business project is in the pipeline.

Tourist infrastructure:

It has a wooded area that can be used for camping (setting up tents), leisure and recreation.

Other aspects of interest: An integrated management of the beach, taking into account its length, could allow the construction of infrastructure related to commerce and coastal tourism.

Intervention proposals:

Promote environmental education among residents and visitors, together with regular cleaning and solid waste collection campaigns by the local population. An integrated beach planning and management plan should be drawn up, divided into sections according to use, taking into account the physical and socio-demographic characteristics of the area. For example, one part could be reserved for fishing and another for fishing.

Images: View of Cuio Beach, presence of campers (tent) and sign of the sea turtle protection program, Kitaganga Project.

-I- Chituca Beach

Location: South of Cuio beach.

Geographical coordinates:

J <u>Far North</u>: South Latitude -13·00' 501" East Longitude -12° 56' 638".

J <u>Far South</u>: South Latitude -13°01' 138" East Longitude - 12° 56' 348".

Length: 1,296 meters.

Width: 12 meters.

Type of beach in terms of use: Non-urban beach with little intensive use.

Type of beach in terms of slope: It has combined characteristics. Up to 15 meters from the northern limit, the beach is of the tombolo type and at the southern end where the fishermen are concentrated, it is of the intermediate type.

Size and color of the sand: Fine brown sand.

Attendance of visitors: Little influx of visitors.

Access routes: By sea.

Environmental status: There is a tendency for solid waste to increase, mainly soft drink and beer cans.

Degree of occupation of the peripheral space: Presence of rustic huts made of sticks, rags and mats and traces of old block constructions.

Tourist infrastructure: There is no tourist infrastructure.

Other aspects of interest: At the northern end, there are high dunes arranged in the shape of a terrace. The beach is not very suitable for bathing, especially for children.

Intervention proposals:

Preserving the beach through environmental education and regular solid waste collection campaigns. Improve the planning of the fishermen's huts, considering that Chituca is one of the beaches with the greatest demographic pressure, due to its proximity to the town of Cuio. Building a road.

Image: Chituca Beach, one of the beaches with the highest population pressure, lacks an access road.

-1- Calohanda Beach

Location: South of the town of Cuio.

Geographical coordinates:

J <u>Far North</u>: South Latitude -13°01' 311"-East Longitude -12° 55' 747".

J <u>Far South</u>: Latitude South -13° 01' 463"- Longitude East - 12°55' 662".

Length: 360.7 meters

Width: 12 meters.

Type of beach in terms of use: Non-urban beach with little intensive use.

Type of beach in terms of slope: Intermediate type beach.

Size and color of the sand: Coarse brown sand.

Attendance of visitors: Few visitors.

Access routes: By sea.

Environmental status: Good state of conservation.

Degree of occupation of the peripheral space: Existence of rustic huts built of sticks, grass and cloths.

Tourist infrastructure: There is no tourist infrastructure.

Other aspects of interest: In the far south you can see detachments of blocks from the limestone rocks.

Intervention proposals: Raise awareness among the population to preserve the beach through environmental education for fishermen's families and study the feasibility of building a land access road.

Image: Calohanda Beach, lacks a land access road.

-2- Tchinhave Beach

Location: South of the town of Cuio.

Geographical coordinates:

J <u>Far North</u>: South Latitude -13°01' 992"- East Longitude -12°55' 050".

J <u>Far South</u>: Latitude East -13° 02' 711"- Longitude East -12°54' 864".

Length: 1,408.5 meters

Width: 25 meters.

Type of beach in terms of use: Non-urban beach with little intensive use.

Beach type in terms of slope: Intermediate **beach** type. However, there is some depth at 3-5 meters from the shoreline.

Size and color of the sand: Coarse, brownish sand.

An influx of visitors: Little influx of visitors.

Access routes: By sea.

Environmental status: Well maintained.

Degree of occupation of the peripheral space: There are a few rustic huts built of sticks.

Tourist infrastructure: No tourist infrastructure.

Other points of interest: At the southern end, there are some cliffs where you can go diving, underwater shelling and other activities. The beach is characterized by its crystal-clear waters.

Intervention proposals:

Reference beach, to be included in the boat trip itinerary. Favorable for implementing tourism projects with a view to boosting quality sun and beach tourism. It is necessary to study the feasibility of building an access road.

Image: Tchinhave Beach, very attractive, but difficult to access and in need of a study to open a road. It could be used for sun and beach tourism.

-1- Noton Beach

Location: South of Cuio.

Geographical coordinates:

J <u>Far North</u>: South Latitude -13° 02' 779"- East Longitude - 12° 54' 559"

J <u>Far South</u>: Latitude South - 13° 03' 20" - Longitude East - 12° 54' 153" **Length:** 1,244.5 meters.

Width: 15 meters.

Type of beach in terms of use: Non-urban beach with little intensive use.

Type of beach in terms of slope: Intermediate type beach.

Size and color of the sand: Fine brown sand.

Visitor turnout: Low turnout.

Access routes: Maritime

Environmental status: Good **environmental** status.

Degree of occupation of the peripheral space: There are a few rustic huts used by artisanal fishermen.

Tourist infrastructure: No tourist infrastructure.

Other aspects of interest: The beach is partly surrounded by hills and cliffs.

Intervention proposals:

Preservation, study the feasibility of building an access road.

Image: Noton beach, crying out for an access road.

-2- Coporolo river mouth

Situated to the north of Praia do Cuio at geographical coordinates: Latitude South: 12° 55' 86" and Longitude East: 12° 57' 801", the mouth of the Coporolo River forms a unique landscape due to its morphology which combines the beach, the estuary and the Lembrança lagoon. This is a splendid place for nature and adventure lovers, which makes it easy to carry out a range of activities related to nature tourism (sun and beach, sports, recreation and scientific education).

Proposals:

Protect peripheral areas through environmental education, create catalogs and leaflets to publicize and promote (marketing) the attraction and place it on the Cuio tourist package route network. It could be used for educational tourism.

Image: Mouth of the Coporolo River, suitable for adventure, sports and educational tourism (geographical hydrography excursions).

Lagoa da Boa Lembranga

Part of the water that does not flow into the sea forms the lagoon with a stationary volume of water depending on the seasons of the year, with a few rainy seasons, usually in the months of March (South

Latitude: 12° 55' 860" and East Longitude: 12° 57' 801"). The lagoon covers a vast area and serves as a source of water supply and cultivation for the surrounding population.

Proposals:

The lagoon and its surroundings can be used for recreational and leisure activities (camping, boat trips, fishing, sports and similar activities). The physical-geographical components of the site allow it to be used for educational tourism, in order to deepen and consolidate the concepts of beach, sand bar, estuary and lagoon, through geographical excursions.

Image: View of the Boa Lembranga Lagoon in its surroundings, a powerhouse for recreational and leisure activities.

2.3 TOURISM POTENTIAL OF LUACHO
4- Mouth of the Tchiuangulula River

The mouth of the Tchiuangulula River is located to the north of the mouth of the Coporolo River, bordered to the north by the seat of Baía Farta, geographical coordinates: South Latitude: 12° 53' 40" and East Longitude: 12° 57' 03.5". It has a coastal morphology with characteristics that resemble the mouth of the Coporolo River, as it also ends in the form of an estuary, in which the presence of sandbars and a lagoon formed by the contact of the waters of the Tchiuangulula River and the Calombe River stand out, forming a landscape worthy of note for good nature lovers, due to its uniqueness (physical-geographical characteristics).

Proposals: Protect the peripheral areas through environmental education, create catalogs and leaflets to promote (marketing) the attraction, in order to place it on the Luacho tourist package route network. It is feasible for recreation, leisure and educational tourism.

Image: The mouth of the Tchiuangulula River, used for recreation, leisure and adventure tourism, sports and teaching.

-1- Lava Lagoon

This small lagoon is formed by the contact of the Tchiuangulula and Calombe rivers before the waters are lost in the sea, thus forming a landscape of tourist interest, as it allows for canoe trips, sport fishing and/or handicrafts. The lagoon[15] runs parallel to the lava beach, which has a volume of water conditioned by rainfall during the rainy season, peaking in the month of March.

  Proposals:
  Encourage the use of the site for recreational and leisure activities (fishing, canoeing and river trips).

---

[15] The Java lagoon is fed by the Tchivangurula and Calombe rivers, forming a unique landscape that will captivate any nature lover.

Image: Lagoa da Lava is a wonderful place for recreation, fishing, canoeing and river trips.

-2- Lava Beach

Location: North of the mouth of the river Coporolo (Cuio).

Geographical coordinates:

J <u>Far North</u>: South Latitude: 12° 53' 28.9" East Longitude -12° 56' 59.4"

J <u>Far South</u>: Latitude South: 12° 55' 86" and Longitude East: 12° 57' 801" (mouth of the Coporolo River).

Length: 4,418 meters

Width: 600 meters.

Type of beach in terms of use: Non-urban beach with little intensive use.

Type of beach in terms of slope: Tombolo beach.

Size and color of the sand: Fine brown sand.

Visitor turnout: Few visitors (artisanal fishermen).

Access routes: By land (from Luacho) and by sea.

Environmental status: Fair, there is scattered plant material on the beach.

Degree of occupation of the peripheral space: There are a few rustic huts used by artisanal fishermen.

Tourist infrastructure: No tourist infrastructure.

Other aspects of interest: The beach with raised sandbars.

Intervention proposals: Preservation of the surrounding area, as the area is suitable for practicing various sporting activities, such as canoeing, mini-surfing and fishing. The site could also be used for geographical excursions, to deepen the concepts of estuary, estuary, beach and lagoon.

Image: Lava Beach, worthy of geographical excursions, deepening the concepts of estuary, estuary, beach.

Image: Sandbanks, reminiscent of the plateau, an excellent place for observing the natural wonders that the area has to offer.

Other natural attractions

In addition to the natural tourist attractions listed above, there are others:

-3- Undembe Lagoon, also known as Canto Lagoon, is conditioned by the flow of the Kuporolo
River. It is inhabited by crocodiles and visited by pink flamingos. It is used for artisanal fishing.

2.2 HISTORICAL AND MONUMENTAL TOURIST RESOURCES

2.2.1. Tourist potential of the headquarters and outskirts

-1-Cabega do Branco

Situated to the north-east of the headquarters, the Cachapéu district[16] is 4 kilometers away, on a 4-minute walk along the Benguela-Dombe Grande road (geographical coordinates: South Latitude: 12° 54' 76.7" and East Longitude: 13° 07' 51.6"). The site has a grave that is an important historical landmark, as oral sources from traditional authorities point out that the grave is that of a Portuguese shitter (1867-1897) who was killed and devoured by a lion, leaving only the head that was buried there to make way for the grave. The site (cemetery) is in a state of disrepair and still needs to be preserved.

Proposal:

Considering the history of the site (cemetery) and the grave of the white shitter in particular, it should be protected and rehabilitated. A security perimeter could be demarcated and the attraction signposted with a descriptive placard that summarizes the essence of the place.

---

[16]This name was given to the neighborhood due to the configuration of an adjacent mountain whose summit is shaped like a hat.

Figure: Campa from the Portuguese cagador (1867-1897) which gave rise to the name "Cabera do Branco". Professor Bandeira, Mr. Lopes (guide) and soba (guide).

-2- English Cambongue Fort

Kambongue Ingles is a historical monument located to the northwest of the town (17 kilometers), a twenty-five minute drive from the turnoff at the entrance to the town and to the right of the Benguela - Dombe road (geographical coordinates: Latitude South: 12° 51'677" and Longitude East: 13° 04' 415"). This important fort was built around the 17th century by English settlers. An imposing work of colonial military architecture, it occupies an area of approximately 1 km^2 , a typical construction of adobe, clay and stone with 80 cm thick walls plastered with clay.

According to the sobas Vivo Maquike and Gongalves, the site served as a military barracks to defend against those coming from the kingdom of Bailundo and was connected to the Catumbela fort. The fort is linked to a triangulation[17,18] with two other buildings, both related to slavery and the colonial reprimand of the Portuguese and British.

Unfortunately, this important heritage site is in a state of continuous degradation due to rain and wind erosion, and is therefore on the verge of complete extinction, which is why it cries out for the merciful hand of the Ministry of Culture to preserve it.[17]

Proposal:

The Communal Administration, the Ministries of Culture and Tourism should join forces to adopt an intervention strategy to improve access (road) to the ruins, in order to facilitate visits to promote and publicize this important site. We should also join forces to attract support and funding to rehabilitate the architectural structure of the heritage site, maintaining its original form (adobes), using specialists in rehabilitation and taking advantage of the local workforce (stonemasons). You could consider marking out a protective perimeter and planting a fence of trees (moringa and acacia). For example, study the feasibility of involving the Teacher Training School (geography and history courses), associations and youth groups (scouts and young people).

Figure: Structure that served as a military administration (left) and area where slaves and/or prisoners were concentrated (right).

---

[17]Tchiúmbua (a dialect name for the ruins) in the Canto neighborhood.
[18]Tchiúmbua, from Bairro do Forno.

Tchiúmbua do Canto (ruins)

The ruins of Canto served as a residence and brandy factory, located in the neighborhood of Canto to the northwest of the headquarters (geographical coordinates: South Latitude: 12° 51' 83.4" and East Longitude: 13° 02' 75.1"). This monument dates from around the 17th century and belonged to a Portuguese family with the surname of Barbosa, who also owned the brandy factory located 15 meters from the residence. This monument is closely related to the English Kambongue, since slavery was prevalent at the time and this family also traded in slaves. It is in a terrible state of preservation, as it is also on the verge of extinction and needs to be preserved.

    Proposal: To study the viability of the site with a view to restoring the heritage and delimiting a protection area and to assess the advisability (profitability) of placing the site on the municipality's tourist route network.

Figures: Ruins (Tchiumbua) of Canto and Residence of the Barbosa Family 17th century.

Tchiúmbua do Forno (ruins)

The ruins of Forno was a residence and brandy factory located in the Bairro do Forno district, to the north of the headquarters at the geographical coordinates: Latitude South: 12° 55' 98" and Longitude East: 13° 05' 67.6". This important ruin, dating from the 16th century, belonged to Mr. Kwekwe at the time, an economic agent in the region. This place is also related to slavery, as the factory workers were slaves who came from the interior of the country, specifically from Bié. This is one of the most degraded ruins, almost completely extinct unless the cultural sector intervenes urgently to preserve and rehabilitate it.

Proposal:

Demarcate an area to protect the ruins and draw up a rehabilitation plan maintaining the architectural features of the heritage. It is important to promote environmental education campaigns for the population, involving the resource to improve the environment and preserve it, in order to place it in the local cultural tourism network.

Figure: "Tchiumbwa" ruins in the Bairro do Forno dating from the 16th century (the state of critical degradation can be seen).

-I- IECA **Church19**

This heritage site is located to the south-east of the Dombe Grande headquarters, in the Calundo neighborhood (geographical coordinates: Latitude South: 12° 58' 17.7" and Longitude East: 13° 05' 99.4"). The religious heritage of the IECA, founded by Canadian and American missionaries in 1942. According to Pastor Dinis Canivete Naval, the church's main figure is Dr. Tuke of American nationality who, together with the Reverend Jesse Chivela Chipenda (Angolan), inaugurated it and left it under the responsibility of the then Pastor Macedo Pina Molongonga.

Pastor Dinis Canivete explains that the symbol of the "cross" was a strategy to avoid persecution by the Portuguese for being a church of American origin, as the Roman Catholic Church dominated at the time.[18]

Proposal:

Considering its historical and symbolic value for the IECA, the church should be rehabilitated, maintaining its original religious architecture. The heritage can be included in the religious and cultural tourism network of the town and the province of Benguela.

Figure: IECA church in Dombe Grande, founded in 1942 by Canadian and American missionaries who promoted evangelization in the area (Bairro do Calundo).

-1- St. Francis Xavier Church

This Roman Catholic Church building is located in the Açucareira neighborhood, at the geographical coordinates: Latitude South: 12° 57' 00.1" and Longitude East: 13° 05' 51.2", built in 1938 and founded in 1960. It is a typical colonial religious architectural heritage site, located in a privileged area of the communal seat of Dombe Grande.

Proposal:

The church must be rehabilitated regularly in order to be preserved, due to its importance to the faithful of the Catholic Church and its symbolic value.

Heritage should be included in the itinerary of the municipality's historical and monumental tourist resources, with a view to boosting religious and cultural tourism.

---

[18] IECA: Evangelical Congregational Church in Angola.

-2- Açucareira 4 de Fevereiro

Ruins of the old and traditional sugar factory closely linked to the history of Dombe Grande, built in 1938, called the 4 de Fevereiro sugar factory, after independence (1975), served as a reference point in the locality, due to its magnitude and contribution to sugar production in Benguela. It was shut down in 1991 and today there is no prospect of it being restored.

Proposal: Part of the factory could be adapted to house the provincial sugar museum. A place where representative samples of old machinery used in sugar factories such as Cassequel in Catumbela and other places in Benguela could be gathered.

Images: Partial view of the sugar factory and symbol of the history of the Dombe Grande sugar factory.

-3- Curráis area

This area is attached to the old sugar industry, which is clearly visible due to the trees that surround it. It is located on the right bank of the main road (Benguela-Dombe Grande). It stands out for having two interconnected places of interest:

a) A kind of museum of the old sugar factory (machinery).

b) A space suitable for recreational and leisure activities, with large secular trees (bongolola)[19] setting the scene.

Proposal: A project could be created inside the corrals to accommodate leisure and recreational activities: a camping area and/or an area with tables and chairs for resting, and possibly investing in a snack bar to take advantage of the shade of the imposing trees in a creative way. It offers conditions for holding cattle fairs and activities related to rural tourism.

---

[19]Imposing trees predominate in many places in the Dombe Grande commune.

Image: Large trees in the backyard of the corrals,

Images : View of old buildings and a sample of old machinery from the 4 de Fevereiro sugar factory.

Dombe Grande Palace

Privileged by its location on the main street of the municipal seat, the palace is a symbolic building of colonial architecture built in 1942, which served as the residence of the former head of the administrative post during the colonial period. It is one of the best preserved references to the colonial administration in the commune.

Proposal: Periodically rehabilitate the building while maintaining its originality so that it continues to be an architectural heritage building, a symbolic reference in the community.

Image: Former palace of Dombe Grande.

Akokoto de Kakolongo

Located on Kakolongo mountain, it is the place where the remains (bodies) of Soba Buekesse (Siwa's predecessor and uncle) and Siwa (Buequesse's nephew) are buried, the latter in 1973. The akokoto is located to the west of the commune's headquarters, a journey of just over 16 minutes at a speed of 40 Km/h. It is a symbolic place for the Mundombes, located on the slopes of the Kakolongo mountain, a karst cave (limestone). Accessing the site is a physical challenge, due to the rocky topography and steep slope of the mountain.

Proposal: Improve the access trail, signpost the route and intervene on the site (cave) with a view to better accommodating and preserving the remains found there. Consider scheduling a specific period for visits that follow traditional rituals[20] with the presence of local sobas. Study the feasibility of including a visit to the site in the tourist itinerary.

Image: View of the mountain where Ackokoto de Kakolongo is located (Image taken)

from the base of the mountain), and the box of the soba Siwa (1973), with its skull in a reasonably preserved state.

Images : Ritual performed by traditional authorities (sobas and elders) before climbing
the mountain where Akokoto de kakolongo is located.

-1-Bridge over the Coporolo River

A landscape (bridge) of interest located 4 kilometers from the entrance to the Dombe Grande commune, with two bridges arranged in parallel, including the narrow old bridge, built in 1972, and the newer, more modern bridge, with two lanes, built in 1912.

Proposal:

Take advantage of the bridges, especially the oldest one, for walking tours of the riverside landscape with groups of visitors during guided tourist excursions.

---

in traditional costume (image). (2) One of the acolytes says a series of words to calm the spirits and ask for permission to visit. Then others take the floor. 3) The wine is poured for all the members of the team and some of the liquid is poured onto the ground. 4. march towards the cave where the Akokoto (body) is.

5. Arrival at the place, the start of another ritual, where one of the elders begins by saying a few words, followed by the distribution of wine and then the elders intervene until the end of the act (this last part lasts about 30 minutes). 6. they leave and meet again at the starting point to say goodbye and thank them for their visit.

Images: South African tourists walking on the old bridge and a partial view of the two bridges over the Coporolo River.

-2- Tcitandalúcua Rock Paintings

Natural resource (heritage) located on the outskirts, to the northeast of the communal seat of Dombe Grande (geographical coordinates: Latitude South: 12° 55' 012"and Longitude East: 013° 10' 088"), occupying an area of 1,200 hectares. The attraction consists of a karst cave that is continually eroded by rainwater during the rainy season. At the entrance, you can see two geometric figures which are called Tchitandalúcua rock paintings.

Proposal:

Improve access, signpost the route and put up a flagpole to make it easier to identify the site. Create an expedition to study the cave in greater depth in order to analyze the possibility of connecting it to the entrance located upstream (South Latitude: 12° 55' 01" and East Longitude: 013° 10' 130") of the main entrance.

Image: Soba Alfredo at the entrance to the Tchitandalúcua cave

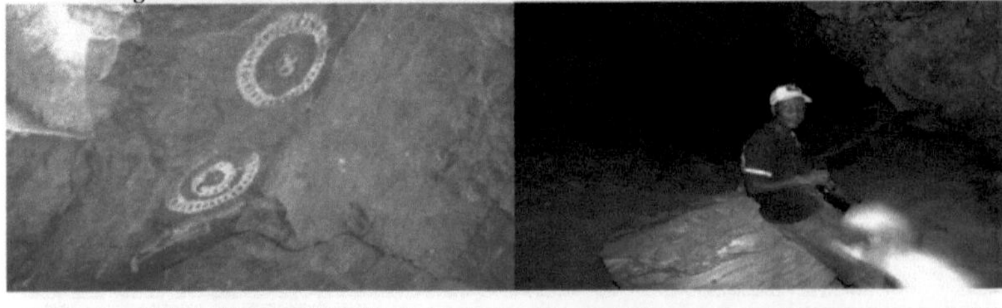

Images: Two circles symbolizing the cave painting, guide Constantine and soba together at the entrance and view from inside the cave.

2.2.2 Luacho's tourism potential
4- Fazenda Santa Theresa (Palm oil factory)

Complex of infrastructures that supported the old palm oil factory in the colonial period, comprising a hospital (1888), office, residential area and agricultural space with palm trees and other crops. The complex belonged to Mrs. Santa Theresa, is currently occupied by the local population and is in a state of disrepair with no prospect of rehabilitation (geographical coordinates: South Latitude: 12° 52' 42" and East Longitude: 013° 00' 39").

Proposal:

Try to protect the infrastructure in the hope that one day there will be a businessman interested in investing in the area and using it for an economically viable agricultural project.

Images: View of the factory, hospital (1888) and residential structure of the farm.

4- Tchiúmbwa Monument (1912-1921)

This landmark monument is located at the geographical coordinates: Latitude South: 12° *54'* 377" and Longitude East: 12° 58' 881", in Luacho, in the Chimbwa district. This work of architecture built from adobe also served as a slave concentration point.

Proposal:

Evict the population living inside the heritage site and mark out a security perimeter to protect the heritage. Subsequently, the building should be rehabilitated, maintaining the original structure, using adobes that can be made in a mobilization campaign guided by a civil engineer.

Images : Views of the heritage ruin in which you can see its adobe architectural features and the degree of degradation that continues.

-I- Visoma Agricultural Farm[24]

Located at the southern end of the municipality, it is an agro-livestock farm that cultivates various crops such as mangoes, vegetables and cereals and has the potential and favorable conditions for tourist activities such as horseback riding, observing some animal species in captivity, including alligators, gibbons, monkeys and other animals.

---

[24]The owner is planning to open a space geared towards tourism, with better conditions to guarantee the life of the animals and allow tourists to visit. He plans to expand and diversify the animal species. To create a kind of zoo capable of attracting more visitors.

Proposal:
Improving the accommodation space for the animals, with a view to ensuring better conditions for tourists interested in observing the captive species on the farm.

Images: Alligators in captivity.

1.1.2. Other monumental historical tourist resources

In addition to the historical and monumental tourist resources presented above, there are other resources in the Dombe Grande commune:

-1- **Sulphur deposit:** exploited by Agucareira, located in Bairro do Ndoloma, north of Dombe Grande.

-2- **Passadeira do Canto: This was** the only access point (entrance) to Dombe Grande until 1972, when the current road came into operation after the construction of the bridge over the Coporolo River.

-3- **Yellow, Red and Blue Composites**: Residential areas built near the Sugar Factory to house the sugar cane plantation workers from other parts of the country, mainly central Angola.

-4- **Akokoto do Kachapéu:** Located in Bairro **do** Ndoloma, 7 km from it and north of Vila do Dombe Grande.

2.3. CULTURAL TOURISM RESOURCES

2.3.1. Cultural Tourist Resources (habits, customs and traditions)

The commune is characterized by a diversity of ethnolinguistic groups with valuable cultural wealth, which is manifested in initiation rituals and the passage to adulthood of boys and girls, weddings, funeral ceremonies, the use of vernacular therapeutics, dangas, languages and endogenous institutions (Gil, 2013).

The Vandombe occupy a prominent place in the preservation of endogenous values, manifested in the practice of rituals, which despite interaction with other cultures and cultural globalization are preserved. These include:

a) **Ekunga or makunga**: A socio-cultural institution for female children and adolescents, it is the female circumcision that used to take place in the months of August-October[25] .

b) **Ekula**: Socio-political institution in traditional authority that precedes the ekwendje;

c) **Ekwednje**[26] : Children's activity, held between August and October;

d) **Otchoto**: Nowadays it appears in the ndombe social context with mere scholastic-religious symbolism.

e) **Ekoya: This** consists of a dialog between the living (usually older people who are prepared for it) and the dead person, which allows them to discover the causes of their death according to the signs they give.

f) **Efeko:** Ritual of female initiation into adulthood. When the girls reach puberty, they are confined to a cubicle, called an otchitenda, for six months, learning everything a woman needs to know. At the end, a big party is held, with all the young men and women in the village taking part in a ceremony where several head of cattle, provided by the girls' parents, are slaughtered. The meat of the slaughtered animals is displayed in a protected place by the girls, who are "armed" with a kind of whip called an etepa, challenging the young men, future suitors, to take the meat they want to roast. When they pass by to pick up the exposed meat, they are struck with the etepa, and only the most daring succeed. It's a ritual that brings young people of both sexes together as a way of getting to know each other, resulting in the formation and preparation of several couples. But there is also the analuongue festival, which is attended by the wealthiest Mundombe shepherds, who offer and slaughter several head of cattle, the meat of which is served to the whole community equally. The wealthiest shepherds thus demonstrate their power through the meat, which they offer to the poorer community at this great feast.

g) **Okuyanga**[27] :   Mundombe funeral ritual,   which precedes the act of
The burial begins at home and ends at the cemetery and lasts an average of 4 hours. It consists of the deceased saying goodbye to family members, friends, etc., in which he (the deceased) communicates with the living through signs or gestures, interpreted by one or more specialists, while the urn carried on a kind of sling, carried by four (4) men, goes round and round the kimbo, accompanied by the deceased's favorite songs. During this farewell to the deceased, questions are asked about authorship/cause of death, succession and guardianship of children, among others. In the absence of the corpse, due to a death that occurred outside the locality, a death is made called **otchikondji** and under these conditions a bit of earth (sand) is extracted from the grave of the deceased and wrapped in a cloth to make okuyanga. According to some of the elders' accounts, this wrapping of earth (sand) has more power than the body itself. This type of funeral ritual is now rare and used to take place at dawn (3 or 4 hours).

Traditional dangas

Among the various dangas existing in the Commune, the most representative can be identified among the mundombes (Gil, 2013):

h) **Osindeta:** Danga exclusive to the Vandombes, also performed by the Vambali.

i) **Ombulunganga:** Fun dance of Mukwandu origin, whose cangao has the same name. It is also practiced by the mukilengues and muhumbi.

j) **Mukeke:** Danga mundombe performed by people linked to magical-religious power. It is performed on various occasions, such as wedding parties, Evamba, Ekula or Okufekala

---

[25]The main reason why most of the events take place between August and October is because this is the harvest period (bonanga), which ensures that food is available within the community.

[26]Ekwendje: Ritual that consists of the passage from male initiation to adulthood (it aims to prepare the young man for the future by assuming his social responsibilities).

[27]Testimonials from Mr. António Cachilingo, Professor and Executive Director of ADSAC (Association for Social Development and Environment).

ceremonies.

k) **Ekoya:** Danga for the elderly.

l) **Otcilombonde:** Danga performed during the celebration of the female rite of passage called Okufekala. It occurs among the Vandombe and Vahanya peoples.

m) **Otchitita: a** song and dance by Vahanya and Vandombe shepherds. For the former, it is performed on the basis of a decision made by the collective of a companionship forged at the School of Initiation (evamba or ekwendje). According to the accounts of the elders, it happens when someone in the community of breeders is possessed by a spirit, at which point a person is elected to perform a ceremony to expel the spirit called otchitamba. It is practiced by cattle breeders. Those who don't have cattle can't sing.

n) **Okambungula and Ehaka:** efiko dangas.

Otchingandji: A person whose danga is ovinganji. The masked dangarino, otchingandji, is a central figure in traditional festivals and rituals, constituting a cultural value that tends to be lost if the necessary measures are not taken to protect it. This traditional dance comes from man's natural impulse, whether it takes the form of jumping for joy or sadness, or the ritual invocation of the gods through a mystical manifestation.

Image: Otchingandji showing off his danga.

Other cultural aspects

In terms of traditional medicine, Dombe Grande is a national reference because it is associated with various facets of traditional medicine, witchcraft and fortune-telling, which is why people from all over the country flock here to solve their problems.

Handicrafts are an integral part of life for the people of Dombe Grande. The objects produced include: Mats, baskets, musical instruments, kitchen objects, ornaments and finials, made from local materials of both plant and animal origin.

Image: Traditional drumming typical of Dombe Grande.

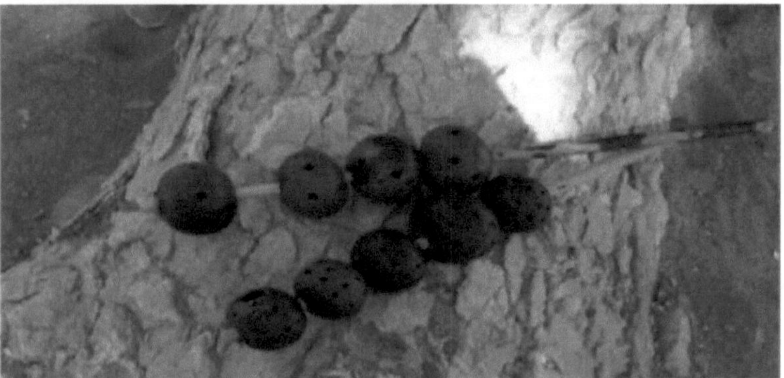

Image: Rattles made with gourds and seeds inside.

Image: Rustic canoe made from bimba logs (Cuio).

Gastronomy is manifested in specific foods and cooking objects. The typical mundombe dish is called "*ondimbe*" and is used to put the conduit in and the "*etamilo*" *is* used to put the píreo in. There is a larger one called "*ohumba*", a kind of bowl in which you can take fuba, corn and all the produce from the fields. The "lombi", the cut branch is called "*endiapata*"; the makunde bean is called "*ongonda*", which is prepared with palm oil. Mundombe prefers coarse fish over other species such as sardines, which are generally not eaten.

Local festivals are linked to events such as *okufekala* 26, *ekwendje* and *ekula, which are* related to the initiation of girls and boys into adulthood. Carnival (February) is also part of the festivities, and is held in accordance with the liturgical calendar[28] [29] . The town has yet to agree on a date for the celebration.

2.4. TOURIST AND SUPPORT INFRASTRUCTURE

2.4.1. Hotel and restaurant services

-1- Hotel A N House

A reference hotel with a category of three (3) stars, located on Rua do Comércio (main road Benguela-Dombe Grande), of modern and recent construction founded in September 2017. The infrastructure has 12 rooms (suites), a restaurant-bar, terrace, swimming pool with relaxation area and a games room with snocker and employs a total of approximately 20 workers to provide related services. The hotel has all the facilities that guarantee a quality stay in a peaceful and safe environment.

---

[28]Okufecala or efiko: The content of the ritual is the same, but the name varies from province to province.
[29]In 2013, the Commune once again took part in the provincial carnival parade with a carnival group, achieving 3rd place.
a

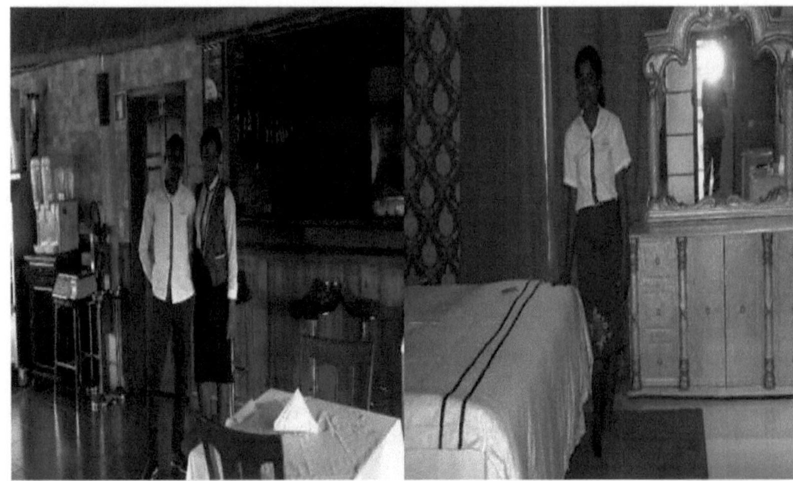

Images: View of the hotel building from the main street, hotel sign, workers and swite

Images: Snocker table in the games room, swimming pool and a painting in the relaxation area, which symbolizes the high taste for art and great style.

-I- Kimbo flavor Esplanade

This uniquely restored infrastructure, built with a wooden base and a grass roof combined with tiles, blends into the typically rural landscape of the area. Sabor do Kimbo (2013) is located just outside the town, on the left-hand side of the road (Benguela-Dombe Grande) towards the province of Namibe. The place is welcoming and very popular with travelers and tourists, mainly from the cities of Benguela and Lobito, who are looking for a cozy atmosphere. The unit has tables, a relaxation area, space for games such as table football and snocker, and there is a fuel pump nearby, making it easy for travelers looking for a rest to continue their journey south or north.

Images: Front view of the Kimbo Esplanade

Images: Young people taking a break, customers and a view of the typical rural landscape (Kimbo).

-1- Eden Bar

Located at one of the crossroads entrances to the main road, not far from the A N House hotel, Eden-Bar is a simple, small establishment, but very popular with young people, as it is situated in a somewhat discreet area, which is why it is so unique. It serves light dishes and snacks accompanied by drinks.

Image: View from the entrance to the snack bar.

-2- Serra da Estrela Esplanade

Located at the entrance to Dombe Grande, before the first bridge towards the town center, this nook stands out for its geographical location, as it welcomes those arriving and bids farewell to those leaving. It offers restaurant services (meals and drinks) to guarantee visitors an excellent stay.

Image: Front view of the esplanade at the entrance to the town.

2.4.2 Other tourist support services and infrastructure

In terms of support services for tourist activities, there is a gas station attached to the Kimbo esplanade, located just outside the town in the direction of Namibe province, and a Banco Internacional de Crédito (BIC) ATM, located on the town's main street.

The town doesn't have a sports hall or a cultural center for young people, as they have few options for recreation and leisure; a lack that can lead young people astray into harmful activities that are detrimental to society.

Inter-municipal access from Benguela and at the internal (communal) level is guaranteed by a few constant buses, Toyota Hiace-type minibuses (Candongueiros) and a few vans of various makes (open-topped) that travel to Benguela and vice versa. These means of transport usually end at the local market, joined by other trucks leaving Lucira (Namibe) for Luanda.

As far as security is concerned, there is a National Police station on the main street, whose mission is to ensure the safety and tranquillity of the residents and visitors, as the town is calm, peaceful and safe.

CHAPTER III

DIAGNOSIS AND ACTION STRATEGIES

3.1. Diagnosis of the local tourism sector

Despite its natural, historical, monumental and, above all, cultural tourism resources, Dombe Grande does not have a strategic plan to boost tourism activities with a view to generating jobs, income and local development, with a positive impact on improving the quality of life of the resident population. On the other hand, the town has little in the way of accommodation, catering, leisure and recreation services, partly due to the timid presence of a few entrepreneurs in the sector who are totally undercapitalized and lack creativity and innovation in the face of the country's current economic situation, which should be seen as an opportunity.

Approaching DAFO analysis

Fortresses	Weaknesses
J Preservation of local culture (local habits, customs and rituals). *J The* existence of numerous valuable natural resources (beaches, rivers, lakes, mountains, flora and fauna and others). *J The* existence of various historical, monumental and cultural tourist resources. *J* Colonial architecture and traces of the slave trade.	*J* Incipient tourist activity. *J There* is no tourism development strategy on the part of the local administration. *J Poor* tourist infrastructure. *J* Lack of specialized professionals in the tourism sector (hotels and restaurants). *J* Decapitalization of the local business community. *J* Poor condition of access roads to tourist sites and attractions.
Opportunities	**Threats**
J Proximity to the province's main cities. *J* Natural areas, monuments and sites. *J The* current importance of cultural, natural, sports and adventure tourism. *J* Current economic crisis. *J* Rehabilitation of the Benguela-Namibe road (Dombe-Lucira).	*J* Trends in environmental degradation. *J A* tendency to lose the values of the local culture. *J* Lack of publicity, promotion and marketing of attractions. *J* Possible entry of foreign investors into the local tourism market.

Flag (2018)

The matrix approximated to the diagnosis presented requires the local authorities of the Administrado Comunal, the Ministry of Tourism (Minotur), the private sector (local and external entrepreneurs) and other tourism stakeholders to take timely action to adopt corrective measures; to make rational use of the *strengths* observed, attacking *weaknesses in* order to minimize them as much as possible, controlling *threats* and taking optimum creative and innovative advantage of the *opportunities that* exist throughout the territory.

The municipality needs to attract foreign investment, which is why it is important to carry out an active tourism advertising and marketing campaign, to make its potential known to the country and the world, while continually creating and improving basic sanitation infrastructure, access roads, improving the water and electricity supply, among other services that support tourism, in order to reduce production costs for entrepreneurs who want to invest in the town.

3.2. Proposals and strategies to boost tourism in the commune

The inventory made it possible to determine that the territory has numerous and valuable natural, historical, monumental and cultural resources (attractions), which make it possible to activate *"tourism in rural areas"* in its various forms. With this in mind, the following proposals are made:

- Design and implement a strategy (program) for publicizing, promoting and *marketing*[30] , to make the tourist potential of the Dombe Grande commune known to others. It is essential that the program is realistic and humble (highlighting only what exists as it is, but looking to the future with sustainability in mind).
- Attracting investment to the municipality to increase and improve the range of accommodation (hotels, pensions and camping), restaurants and other services, because the absence of these structures makes the area a transit point;
- Improve accessibility (roads) to the key locations: Cuio; Luacho and other areas with tourist potential;
- Create public-private cooperation projects, because in the local and current context this is crucial to the success of actions in tourism planning and management;
- Draw up an agenda (calendar) of municipal entertainment activities that includes natural and cultural attractions and folklore, in conjunction with tour operators and national agencies;
- Create natural interpretation programs, where all natural resources will be covered. Prioritize the design of thematic hiking trails and their signposting throughout the territory;
- Investing in environmental education, awareness-raising and the preservation of the "*natural, historical-monumental* and *cultural heritage*" involving the population, residents, tourists and tourism operators.
- Investing heavily in improving the supply of electricity, water and basic sanitation, increasing household connections.
- In conjunction with traditional authorities and local churches, draw up a calendar of cultural, religious and local events;

The strategic proposals set out above require coordination on the part of the different actors and stakeholders in the local tourism system, so that the plans reflect the active and dynamic participation of everyone, in order to avoid exclusion so that the plans (programs) are coherent and functional, because most of the time the plans (programs) are a failure because they are drawn up by outside experts without knowledge of the local reality and without consulting and listening to the opinion of the citizens.

---

28 **SPECIFIC** ANNEX: *Strategies for publicizing, advertising, promoting and marketing tourism in the Dombe Grande commune.*

FINAL CONCLUSIONS

The result of the inventory of tourism resources carried out and the analysis of internal and external factors using the methodology of the analysis of Weaknesses, Threats, Strengths and Opportunities (DAFO), it can be clearly and unequivocally concluded that the municipality of Dombe Grande has a vast and diversified potential of tourism resources (places and attractions), many of them of reference due to their uniqueness, which can be enhanced with a view to their valorization and economic profitability as competitive tourism products.

A total of 15 (fifteen) natural tourist resources, 16 (sixteen) historical-monumental tourist resources, 14 (fourteen) cultural resources and 4 (four) tourist infrastructures were recorded: 1 (one) restaurant, 1 (one) hotel and 1 (one) esplanade. Despite the tourism potential registered, it can be said that the development of tourism in the commune is incipient (very low).

The administration and Minhotur do not promote tourist activities, nor is there any strategy (plan) to boost this important sector of the economy, which, if it is organized and planned involving public management and private initiative, could become a strategic sector that complements the predominant agricultural activity in the area.

Dombe Grande has sufficient resources to attract visitors (tourists), as it has *historical and monumental* tourist resources of varying degrees of uniqueness, some of which are in an acceptable state of preservation and which could enable the creation of a sugar museum, for example (agucareira). Above all, it has a set of habits, traditions and customs, reflecting the ethnolinguistic groups that inhabit the territory, which makes it possible to boost cultural tourism by leveraging manifestations such as traditional dangas, handicrafts (basketry), gastronomy, the rite of passage into young-adult life; the "efiko-ekwendje" among other cultural manifestations that nevertheless retain ancestral traits. The town also has valuable *natural tourism* resources that can be used to boost nature tourism, such as the natural landscapes, rivers, hills, desert and virgin beaches which, together with the rural nature of the town, can result in a diversified range of quality tourism products and services and become a natural-cultural destination of reference in Benguela.

It is important to encourage the local and foreign business community to invest in the area and to implement a program to publicize, promote and market tourism, on the one hand, and to seriously improve access (roads, Cuio, Luacho and others), accommodation (lodging and restaurants) and signposting of attractions, on the other, as fundamental elements for a successful local tourism development strategy.

BIBLIOGRAPHICAL REFERENCES

Administrado comunal do Dombe Grande (2017): Annual report.

Alexanda Aparicio, M. (S/D): "Dombe Grande a implantado dos dízimos e suas consequencias na regiao os conflictos militares 1852-1859". Summary of local history.

Bandeira, M. F. *et. al;* (2011): "Inventory and diagnosis of the tourist potential of the coastal municipalities of the province of Benguela". Provincial Directorate of Hotels and Tourism. Benguela.

Crosby, A. (Editor) (2009): *Re-inentando el turismo rural*. Editorial LAERTES.

Díaz, B. (2011): *Diseño de productos turísticos.* Editorial Síntesis, S. A. Madrid.

Ejarque, J. (2003): *Destinos turísticos de éxito. Design, creation, management and marketing.* Ediciones Pirámides (Grupo Anaya, S. A.), Madrid.

Gil Tomás, I. (2013): "Analysis of the historical-monumental and cultural tourist potential of the Dombe-Grande Commune". Final paper for the degree of Licentiate in Education in the specialty of Geography, Department of Natural Sciences, ISCED Benguela.

Kajibanga, C. M. (2009): *Coreografía rural-Uma contribuicao para o estudo sociocultural de Benguela*" KAT-Empreendimentos e Consultoria, Benguela.

Leno Cerro, F. (1991): "Los recursos turisticos en un proceso de planificación: inventario y evaluación". *Papers de turisme*, 7:7-23.

Ministry of Tourism Industry and Energy (2005): "Inventory of tourist resources in the municipality of Sao Lourengo dos Órgaos". Santiago, Cape Verde.

<h1 align="center">ANNEXES</h1>

SOURCES OF INFORMATION AND ORAL DATA

N°	NAME	FUNCTION
01	Bertolino Asdruba Soares Morais	Communal Administrator of Dombe Grande
02	Joâo Pedro Paulo Moma	Communal Governor of Dombe Grande
03	Martinho Lopes	Head of Culture - Dombe Grande
04	António César Machado	Soba do Cuio
05	Feranando Joaquim Tchivolongo	Ndolova Neighborhood Soba
06	Abílio Jongolo Kafundanda	Soba from the Altumbo neighborhood
07	Cambole Tchindula	Soba from the Tchiculututo neighborhood
08	António Machado	Kasseque Neighborhood Soba
09	Constantine	Acting head of inspector

10	Alberto António	Soba do Canguengo
11	António Alfredo	Coordinator of Canguengo/ Adj. of Soba
12	José Manuel	Luacho's Soba
13	José Tchitumbo	Luacho's deputy soba
14	Gonsálves Paulo "Kapalandanda"	Soba from Bairro do Forno

I want morebooks!

Buy your books fast and straightforward online - at one of world's fastest growing online book stores! Environmentally sound due to Print-on-Demand technologies.

Buy your books online at
www.morebooks.shop

Kaufen Sie Ihre Bücher schnell und unkompliziert online – auf einer der am schnellsten wachsenden Buchhandelsplattformen weltweit! Dank Print-On-Demand umwelt- und ressourcenschonend produziert.

Bücher schneller online kaufen
www.morebooks.shop

info@omniscriptum.com
www.omniscriptum.com

Printed by Books on Demand GmbH, Norderstedt / Germany